Star Atlas

Edited and introduced by **Dr Jacqueline Mitton and Dr Simon Mitton**

Contents

Jonathan Cape
Thirty Bedford Square London

STAR ATLAS
First published in Great Britain by Jonathan Cape Limited, 1979

Created, designed and produced by Trewin Copplestone Publishing Ltd, London

Phototypeset by SX Composing Ltd, Rayleigh, Essex
Printed in Italy by New Interlitho SpA

British Library Cataloguing in Publication Data
Mitton, Simon
  Star atlas.
  1. Stars – Atlases
  I. Title   II. Mitton, Jacqueline
  523.8'903   QB65
  ISBN 0–224–01715–2

Mapping the starry skies

On a clear, dark night, far from the distracting glare of city lights, you might see as many as two thousand stars. Against the silent backcloth of the universe they seem almost close enough to touch. In addition to the individual stars, the silvery band of light known as the Milky Way spans an arc of the heavens. Over the whole sky, in both northern and southern hemispheres, your eyes can see nearly a thousand stars without difficulty and over six thousand when fully adapted to the darkness. The number of visible stars increases dramatically when you use even simple instruments, rising to thirty thousand with small field glasses and hundreds of thousands with an amateur telescope. The *Star Atlas* shows the stars that are visible to the naked eye, together with the star clusters, nebulae and galaxies that can be seen through a small telescope or field glasses.

The distance to the objects that we are interested in are immense compared with any terrestrial standard. Even the very nearest star is forty million million kilometres from us. In fact, all the stars are so far away that, for the purposes of mapping their positions in our sky, we may imagine that they are all at the same distance. In reality they are at different distances, but these differences do not affect the appearance of the star patterns in our sky.

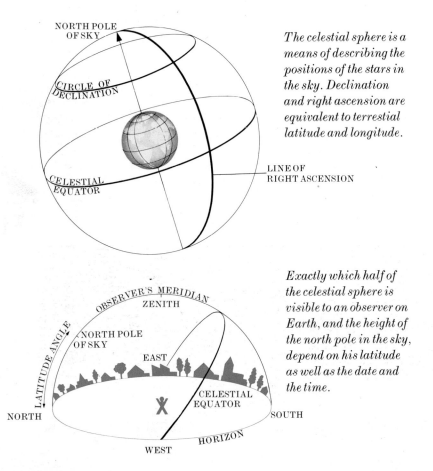

The celestial sphere is a means of describing the positions of the stars in the sky. Declination and right ascension are equivalent to terrestial latitude and longitude.

Exactly which half of the celestial sphere is visible to an observer on Earth, and the height of the north pole in the sky, depend on his latitude as well as the date and the time.

The celestial sphere

In ancient times the astronomers imagined that the stars were fixed to a transparent sphere enclosing the Earth, a concept still used today to make a map of the sky. It is on this imaginary transparent shell of stars, termed the celestial sphere, that we set out the reference system used to record star positions.

A reference, or coordinate, system used in map-making enables users to locate places of interest rapidly. Some atlases, for example, divide each page into squares that are described by a letter of the alphabet and a numeral (e.g. J7). The standard reference system for terrestrial maps is based on latitude and longitude. Latitude is a measure of position north or south of the Earth's equator and longitude is a measure of position east or west of the prime meridian, a standard reference line that passes through Greenwich, England. Just two such quantities uniquely describe the location of any point on the surface of the Earth.

In the astronomical grid system the coordinate that is similar to longitude is called right ascension, and the coordinate corresponding to latitude is called declination. By specifying the right ascension and declination of a star, we can fix its position precisely on a map. Right ascension is measured west of the point in the sky where the Sun happens to be on 21 March (the spring equinox); this point is the astronomer's equivalent of the geographer's Greenwich. Declination is the angular distance of an object from the equator of the celestial sphere. The circles of right ascension intersect at the north and south celestial poles, whereas the circles of declination are parallel above and below the celestial equator.

On the star maps in this atlas the right ascensions are marked round the periphery of the discs (Maps 1 and 2) and along the top and bottom arcs of the segments (Maps 3 to 8) in hours. The complete circuit round the sky is divided into twenty-four hours, each subdivided into minutes and seconds, rather than into 360°. This system, established long before the invention of electronic calculators, makes it easier to relate star positions to the sidereal time shown by astronomical clocks. Declination is marked in degrees along the sides of the segments.

From the surface of the Earth an observer can see only half of the celestial sphere at any one time. The precise view depends on the latitude of the observer. An observer at the Earth's north pole will see the celestial north pole immediately overhead, whereas an observer further south sees the celestial north pole at a lower angle in the sky. An observer in the northern hemisphere can never see objects that have a declination more than 90 degrees south of his latitude. Similarly, for an observer in the southern hemisphere objects that have declinations more than 90 degrees north of his latitude are always invisible. For example, an observer at latitude 50°N cannot see the stars below declination −40° (i.e. 40°S), and an observer at latitude 20°S cannot see stars above declination +70° (i.e. 70°N).

Opposite : The changing aspect of the night sky – Orion is shown at the same time in the evening at intervals of one month.

Choice of the correct maps

A valuable concept in using the maps is that of the observer's meridian. This is an imaginary semicircle on the sky from the north point of the horizon, through the highest point of the heavens, to the south point of the horizon.

Because the Earth is rotating, the starry dome of the sky appears to us to change its aspect in the course of the night. The Earth takes 23 hours 56 minutes to make one rotation relative to the distant stars. Since this is four minutes less than an ordinary day, a given star will rise four minutes earlier, and cross the observer's meridian four minutes earlier, on successive days. This change in the sky's appearance is so slight from one night to the next that you must look carefully to notice it. However, over the course of a month the change in the sky's appearance at the same hour of night is quite apparent. Even more marked is the change from season to season, the winter sky being very different from the summer sky.

The rotation of the Earth and the fact that this rotation takes slightly less than one day mean that the choice of the correct map depends on the local time and the date. Each map is accompanied by a table that shows the local times and dates at which its central right ascension line coincides with the observer's meridian. The map corresponding to the observer's meridian shows the view to the immediate south of an observer in the northern hemisphere, and to the immediate north of an observer in the southern hemisphere. Adjacent maps show the stars to the west and east.

Circumpolar stars

At most latitudes of observation some star groups never set. They are known as circumpolar stars. A star is circumpolar at a particular latitude if its declination angle plus the latitude angle exceed 90 degrees. For example, the stars of the Big Dipper or Plough in Ursa Major (declination $+50°$) are circumpolar at all latitudes above 40°N. Similarly, Crux Australis, the Southern Cross (declination $-60°$), is circumpolar throughout Australia, all parts of which are south of latitude 30°S. At the north pole all stars north of declination 0° are circumpolar. At the equator almost all stars can be seen at some time and no star is circumpolar.

Maps 1 and 2 show stars that are circumpolar at high northern and southern latitudes.

As the latitude of the observer changes, the stars that are visible will alter. An observer in Europe or North America travelling south will notice Ursa Major becoming lower in the sky, and new southern stars coming into view. At the Earth's equator Polaris, the Pole Star or North Star, lies right on the horizon and the celestial equator passes overhead.

Magnitude

The apparent magnitudes of stars are indicated on the maps according to a tradition twenty-two centuries old. For a casual observer the most obvious way in which stars differ is in their brightness as perceived by the human eye. The Greek philosopher Hipparchus devised the system still in use today. He called the brightest stars first magnitude, the dimmest sixth magnitude, and assigned the four intermediate magnitudes to stars between these extremes. A difference of one magnitude between two stars corresponds to a difference of 2.5 times the amount of light energy received at the Earth. Modern measuring instruments are capable of detecting very small differences in star magnitudes, so these are sometimes quoted to the nearest one-hundredth of a magnitude interval. In this atlas the star magnitudes are indicated to within one-tenth of a magnitude by the size of the printed star image. A magnifying eyepiece with an engraved scale is needed to read the magnitudes to this accuracy, but they can be estimated to within about half a magnitude by means of the key that accompanies each map.

The Milky Way

The blue-toned areas that appear on the maps indicate the hazy boundaries of the Milky Way. This beautiful structure in the night sky is due to the combined effect of the light from thousands of millions of very faint stars, individually much too feeble to be seen distinctly by the unaided human eye. But, as Galileo discovered in the seventeenth century, even a small telescope or field glasses will resolve this silvery mist into uncountable crowds of stars. The richest star fields are found in Sagittarius, the constellation that contains the central part of our Galaxy.

The constellations

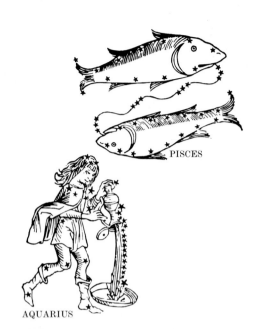

PISCES

ARIES

TAURUS

GEMINI

CANCER

AQUARIUS

A striking feature of the night sky is that the stars seem grouped into patterns. From the beginning of time this must have impressed watchers of the skies, and the practice of assigning names to particular star groups arose in prehistory. Certain star patterns, such as the Plough or Big Dipper, have been considered as distinct patterns since the very dawn of astronomy.

Constellations were initially conceived as star patterns that were thought to resemble the objects, animals, and heroes of popular mythology. Many of the names for northern hemisphere constellations are Greek in origin, with some Babylonian contributions. These classical constellations are almost a condensed account of the major Greek myths, and they were probably used as a natural picture gallery by the myth-tellers.

The Greek star catalogues listed forty-eight constellations, including, of course, the twelve constellations in the zodiac. In ancient times zodiacal constellations were regarded as being of supreme importance because people believed that the Sun made an annual journey round the heavens and passed through each of these constellations on its way.

In the seventeenth century the hitherto uncharted southern skies were mapped and several new constellations were added to the classical list. By the middle of the eighteenth century all constellations had received their present names, but they had no well-defined boundaries. Stars near the borderline of a figure would be

Part of our Milky Way in Cygnus shows the
North America Nebula (top left) and
Cygnus Loop (lower left).

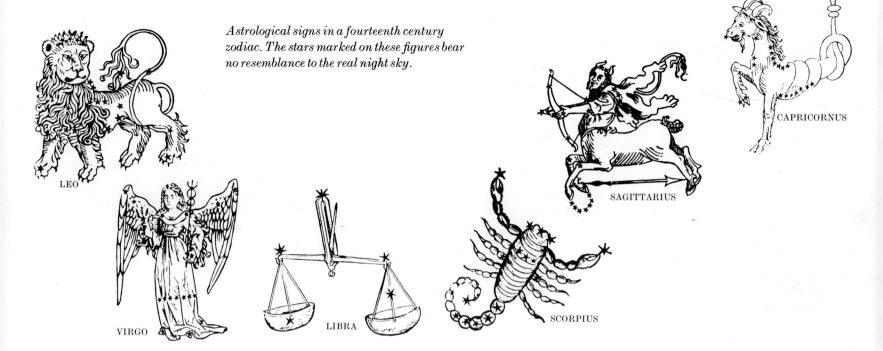

Astrological signs in a fourteenth century zodiac. The stars marked on these figures bear no resemblance to the real night sky.

LEO

VIRGO

LIBRA

SCORPIUS

SAGITTARIUS

CAPRICORNUS

assigned at random to two constellations, which caused much confusion. Finally, in 1925, the International Astronomical Union fixed the boundaries of all the 88 constellations by running the limits to each area along the lines of right ascension and the parallels of declination. This scientific definition of the zones follows the traditional outlines as closely as possible, while ensuring that any star can definitely be assigned to its proper constellation. The international system of constellation boundaries is shown by broken lines on the maps in this atlas.

The constellations have a variety of popular names, as well as internationally recognized Latin names. Astronomers often abbreviate the Latin name to three letters, and it is standard practice to use the Latin genitive case (e.g. Ursae Majoris is the genitive case of Ursa Major, the Great Bear) when attaching a constellation label to a star name (e.g. Alpha Ursae Majoris for the brightest star in the Great Bear). The Latin name, abbreviation, genitive case, popular name, and map reference number of all the constellations are listed in the table on page 30. In order to make the maps clear and easy to read, only the Latin name is marked on them. As the table shows, many of the new constellations of the southern skies have unromantic names, such as the Air Pump and the Clock, a reflection of the technology of the time when they were named.

Asterisms

The prominent star patterns within constellations are known as asterisms. Two well-known asterisms are the Pleiades, or Seven Sisters, in Taurus and, of course, the Plough, or Big Dipper, in Ursa Major. The latter is popularly assumed to be a constellation, whereas it is merely the hindquarters and tail of the Great Bear. Two stars in

the Plough or Big Dipper, known as the Pointers, accurately mark out the direction to Polaris and are useful in locating it quickly. In parts of England this asterism is still called "Charles' Wain", as it was in Shakespeare's day, and the identification of the stars with a wagon or long chariot stretches back to Babylonian times.

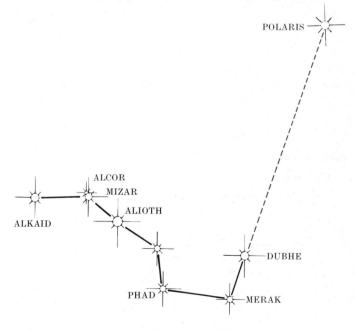

POLARIS

ALCOR
MIZAR
ALIOTH
ALKAID
DUBHE
PHAD
MERAK

Merak and Dubhe in the Plough/Big Dipper are known as the Pointers because the line joining them leads on almost directly to Polaris.

Star names

The names of the brightest stars show the classical and medieval Arabic influence, but hardly any of these beautiful names are regularly used by professional astronomers. The names of the few dozen brightest stars, such as Sirius in Canis Major, Procyon in Canis Minor, and Canopus in Carina, are still used, but proper names are inconvenient when dealing with the thousands of familiar stars visible to the eye. The most convenient system for identifying objects visible to the naked eye was devised in 1603 by Bayer. This system designates the brighter stars in a particular constellation with the letters of the Greek alphabet: α, β, γ, δ, etc. Generally, the letters are assigned in descending order of brightness, with α denoting the star of brightest appearance, β the second brightest, and so on. Sometimes, however, the lettering follows the shape or sequence within a constellation instead of the relative brightness; thus in Ursa Major the seven stars of the Plough/Big Dipper are α, β, γ, δ, ε, ζ, and η, where δ is the faintest of these. In other case, the giant southern constellation Argo Navis was divided into three constellations, Vela, Carina, and Pyxis, without renaming the stars of Vela and Carina, so that Vela has been left with no stars designated α or β.

The seven stars that make up the Plough/Big Dipper have historic Arabic names and Greek letter designations. Phad is also known as Phecda, and Alkaid as Benetnasch.

MIZAR ζ

ALIOTH ε

DUBHE α

ALKAID η

MEGREZ δ

MERAK β

PHAD γ

Star catalogues

When the Greek alphabet did not suffice, lower and then upper case Roman letters were used instead, as far as the capital letter Q. The Roman letters are primarily encountered in the southern constellations. Fainter stars are identified by means of their number in a star catalogue. In 1725 a catalogue prepared by John Flamsteed, the first Astronomer Royal, was published posthumously. It gave the stars within a constellation the numerals 1, 2, 3, 4, etc. roughly in order of right ascension. One consequence of this is that many of the brighter stars have a letter (Greek or Roman) as well as a Flamsteed number. In this star atlas most stars brighter than apparent magnitude 5.5 are labelled with their letter or by Flamsteed number if they have no letter. For most elementary practical purposes this is sufficient. Professional star listings, such as the Catalog of the Smithsonian Astrophysical Observatory (on which the maps in this atlas are based), generally list the stars in order of right ascension without regard to constellation. All stars on the star maps have a six-digit code in the SAO Catalog used by professional astronomers, even though they might also have other names or numbers. Because there are various professional catalogues, to avoid confusion these codes are not printed on star maps. For most amateur observers the Bayer or Flamsteed designations, which are printed on the maps, are sufficient.

The nomenclature of stars with variable brightness is also somewhat complicated, being governed by tradition rather than science. In the nineteenth century Friedrich Argelander labelled those variable stars that were not already designated by a Greek or Roman letter. He began at the letter R (since in one constellation letters as far as Q had already been allocated) and assigned the letters from R to Z to the conspicuous, unnamed, variable stars in each constellation. After Z he adopted the double form RR to RZ, SS to SZ, and so on to ZZ, giving up to 54 variables in any one constellation. Thereafter, AA to AZ, BB to BZ, etc., was employed, omitting J, to give a total of 334 combinations. All subsequent variables after QZ are given the letter V followed by numerals.

The complete proper name of a star includes the Latin genitive case form of the constellation name: α Ursae Majoris, R Coronae Borealis, 32 Virginis, g Herculis, V1500 Cygni, RR Lyrae, etc.

The Greek alphabet

The lower case Greek alphabet is used to designate the bright stars within the constellations.

α	alpha	ι	iota	ρ	rho
β	beta	κ	kappa	σ	sigma
γ	gamma	λ	lambda	τ	tau
δ	delta	μ	mu	υ	upsilon
ε	epsilon	ν	nu	ϕ	phi
ζ	zeta	ξ	xi	χ	chi
η	eta	o	omicron	ψ	psi
θ	theta	π	pi	ω	omega

Left: Open cluster in Cancer, M67; Right: Globular cluster in Hercules, M13; Opposite: Trifid Nebula in Sagittarius, M20

Clusters, nebulae and galaxies

The Great Nebula in Orion, M42

There are many interesting objects apart from solitary stars to see in the night sky. In centuries past astronomers termed many hazy patches of light 'nebulae', the Latin word for clouds. The most prominent nebulae visible to the unaided eye are the Andromeda Nebula (for northern observers only), the Orion Nebula, and the Magellanic Clouds (for southern observers only).

The Messier and NGC catalogues

Charles Messier, the eighteenth century French comet hunter, drew up the first rudimentary listing of the nebulae in order to keep track of the fuzzy but permanent fixtures that kept confusing his searches for comets. Messier's list of nebulae is not in any sense a systematic catalogue, but it does give descriptions and positions for about one hundred of the prominent nebulae in the northern sky. His numerical designations are still in regular use for most of the nebulae that can be seen with small telescopes. The objects noted by Messier are assigned his own reference number, prefixed by the letter M. Thus M31 is the Andromeda Nebula and M42 the Orion Nebula. Messier's list includes objects that are difficult to see in small telescopes. Therefore, although all the objects recorded by Messier are marked by M-number on the maps in this star atlas, the list and description of nebulae on page 31 includes only those Messier objects suitable for viewing with small instruments.

The main professional catalogue of nebulae is the New General Catalogue drawn up by J.L.E.Dreyer in 1888, and revised twice since then. Objects in this catalogue are referred to by their number (up to four digits) preceded by NGC. Certain prominent NGC nebulae are indicated on the star maps by a nebular symbol (☸) and the NGC number, although the 'NGC' prefix has been omitted. In the southern skies only NGC numbers are quoted because Messier listed nothing south of declination −35°. In a few cases nebulae have names that follow the same convention as the nomenclature for normal stars, for example, 47 Tucanae, κ Crucis and h Persei.

Nebulae

The list on page 31 gives the catalogue number, celestial position, map reference and physical nature of the brighter nebulae suitable for viewing with a small instrument. The approximate magnitude of the nebula is also indicated, but this is difficult to evaluate for an extended source of diffuse light. Because the faint light from a nebula is spread over an area of the sky it appears in an instrument as a soft glow, almost ghostly in many cases. The effect is subtle and repays careful and sustained viewing, as the eye slowly recognizes the delicate structures. Some idea of the subjective element involved in viewing nebulae may be gained from the fact that even the great observer Sir William Herschel incorrectly concluded that many of these mists showed marked variations in their structure with time. The more evasive nebulae are best detected by sweeping the instrument slowly through the field. Many nebulae, however, can be enjoyed without any difficulty, especially the compact globular clusters. All nebulae will be seen at their finest on dark, clear nights with no moonlight or nearby artificial lights.

The eighteenth and nineteenth century astronomers used the term 'nebula' indiscriminately. Now, astronomers distinguish between clouds of glowing gas, clusters of stars, and great galaxies far beyond the Milky Way. These distinctions are included in the list of nebulae. The term 'nebula' is now reserved for objects that truly are made of gas and dust. The most spectacular viewing is provided by the emission nebulae, which are glowing clouds of gas with internal temperatures of 10 000–20 000 degrees Centigrade. At such a temperature the gas of the nebula (hydrogen, oxygen, and nitrogen, for example) glows just as it does in the electrical discharge lamps used for street lighting and advertising signs. Without doubt the Orion Nebula (M42) is the finest object of this type, being beautiful in field glasses and spectacular in a telescope. It glows with a sea-green sheen and structural features can be discerned on a really dark night.

Galactic nebulae

Planetary nebulae were so named because their spherical shape causes them to appear like the disc of a planet when seen with a small telescope. They are a form of highly-evolved, or old age, star in which the outer atmospheric layers have been blasted into space to form a surrounding shell of glowing gas. In a few cases you can see the intensely hot (100 000 degrees Centigrade) central star as a point of light in the centre of the nebula. Two planetaries suitable for viewing with small telescopes, but not binoculars, are the Dumbbell (M27) in Vulpecula and the Ring (M57) in Lyra.

Other types of nebula include the dark nebulae, which are clouds of cold gas and dust, and reflection nebulae, which are seen merely because they reflect light from stars that are near them. There are no examples of either type that can be seen through small telescopes. The well-known Horsehead Nebula is a dark nebula, but it is only brought out against the background light by photography.

The nebulae provide the keen sky-watcher with many opportunities to record deep sky wonders by photography. In order to capture the glories of the nebulae it is essential to be able to make time exposures of at least several minutes. Consequently, a telescope with a rock-firm mount and an electric drive is essential. The camera should be a single-lens reflex; adaptors can be purchased to fit the camera body to the telescope. By experimenting you can find the best exposure times and film types. Astrophotography is a rapidly expanding aspect of amateur astronomy: the equipment is not prohibitively expensive and the rewards are considerable.

Among the objects within the Milky Way, supernova remnants deserve a mention. These are the exploded wrecks of dead stars, the Crab Nebula (M1) in Taurus being a well-known example. This is a difficult object for an unskilled observer to see: a very dark and clear night is essential.

Dumbbell Nebula in Vulpecula, M27

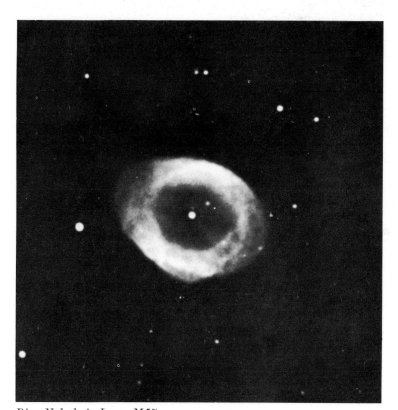

Ring Nebula in Lyra, M57

Clusters of stars

Star clusters are families of stars that may consist of anything from a few dozen to perhaps a million members. Two broad classes exist: open clusters and globular clusters.

Open clusters are loosely structured groups of stars, typically with a hundred or so members. They are mostly located in the plane of the Milky Way. Open clusters are relatively young, astronomically speaking, because most of them formed in the last 100 million years, and some are even younger than this. New stars condense inside cosmic gas clouds and the family of young stars that is spawned stays clustered together initially, forming an open cluster. The finest ones in the night sky include the Pleiades (M45), h and χ Persei (NGC 884 and 869), κ Crucis (NGC 4755), and M7 in Scorpius. The Pleiades is one of the youngest clusters known, having condensed about ten million years ago. One popular name of this glowing patch of light is 'the Seven Sisters'; although most people can only see six stars with ease, binoculars will show several dozen, and the total membership is over two hundred. Between the Pleiades and Orion lies the Hyades cluster, but this is too sparsely scattered over the night sky to be impressive.

Globular clusters are very ancient, perhaps ten thousand million years old. Visually, they are compact spheres of stars. Even in small instruments the globular clusters look like soft spheres of light, although an enlarged photograph taken with a professional telescope will show individual stars at the periphery of the cluster. Globular clusters were formed before the Milky Way collapsed to a disc, and they are now mostly lonely families thousands of light years away from the plane of the Milky Way. Two prominent globular clusters in southern skies, 47 Tucanae and ω Centauri can be seen by the naked eye. Good ones to look for in the northern sky are M13 and M92 in Hercules and M15 in Pegasus, all of which require field glasses.

Globular cluster in Canes Venatici, M3

Globular cluster in Centaurus

Open cluster, the Jewel Box, in Crux

Double open cluster in Perseus, h and χ Persei

The Pleiades in Taurus, M45

Galaxies

Our Galaxy, with its clusters and nebulae, measures one hundred thousand light years from end to end. Far beyond this starry spiral, which we see from the inside as the Milky Way, the extragalactic nebulae, or galaxies, are found. These are huge star families containing as many as a million million stars. The Magellanic Clouds (declination −70°S, map 2) are the nearest galaxies; they are dwarf systems, satellites of the Milky Way, almost 200 000 light years away. Both are conspicuous clouds of light to the naked eye and splendid sights when viewed through an instrument. Within the Large Magellanic Cloud (LMC) the Looped Nebula (NGC 2070) is visible to the naked eye.

Northern hemisphere observers can just see the great Andromeda Nebula (M31) with the unaided eye. At two million light years distance from us, this is the most remote object visible to the naked eye. Binoculars will reveal a soft elliptical glow. M31 is an example of a spiral galaxy, in which the bright young stars are concentrated in two spiral arms wrapped round the galaxy. However, it is not possible to see the spiral arms of such galaxies through small instruments. The beautiful photographs of spirals reproduced in many books are made by time exposure, and they show a wealth of detail that cannot be discerned by the eye and telescope. This is because the eye, unlike a photographic plate, cannot store the light received to build up an image of faint structure. Consequently, small telescopes and purely visual techniques will normally reveal only the bright nuclear regions of galaxies.

Among the brighter spiral galaxies to search for are M33 in Triangulum, a 'face-on' spiral, and M81 and M101 in Ursa Major. The galaxy M51, in Canes Venatici, is the famous 'Whirlpool Nebula'; its spiral structure was first discerned by the third Earl of Rosse in 1848, using a gigantic 72-inch reflector telescope, which he made in 1845. The spiral itself cannot be seen with a small telescope, although the central regions of the galaxy are visible.

Elliptical galaxies are lemon- or cigar-shaped, but all of them are fainter than apparent magnitude 9, and they do not feature in the table on page 31 because they are not visible through field glasses. But the following elliptical galaxies can be seen with more powerful optical instruments and are marked on the maps: M49 (an easier one), M60, M86, M87, and M89, all in Virgo; and M32, the companion to M31, in Andromeda.

Among the jumble of excited and irregular galaxies, two that are visible in a medium-sized telescope appear in Messier's catalogue and are indicated on the maps. M77 in Cetus is the well-known Seyfert galaxy NGC 1068, fifty million light years away, which has an intensely active central region. Its apparent magnitude is about 9. The irregular galaxy M82 in Ursa Major is a strong source of cosmic radio waves. It is possible that this galaxy has collided with an isolated cloud of intergalactic gas.

The Large Magellanic Cloud

The Small Magellanic Cloud

Spiral galaxy in Triangulum, M33

Spiral galaxy in Andromeda, M31

M82 and M81 galaxies

Whirlpool galaxy, M51

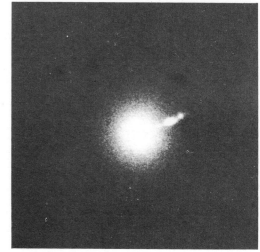

Central part of M87

Variable and double stars

Some stars vary in magnitude, and these variable stars are of several physical types. Intrinsic variables fluctuate because of physical changes in the stars. The most famous of these are the Cepheid variables, which are named after the prototype δ Cephei. This star ranges between visual magnitudes 3.6 and 4.3 over a period of 5.37 days. Cepheid variables alternately swell and shrink, and these changes in radius cause corresponding changes in the radiant energy. The Cepheids are extremely regular, predictable, variable stars.

Long-period variables take from three months to three years to run through one cycle of the variation. The spectacular star Mira Ceti, or o Ceti, has a period of 331 days. It fluctuates between magnitudes 1.7 and 10.1 and is visible to the naked eye for about eighteen weeks in each cycle. Mira is a red giant star that swells up and subsequently contracts. Other red giant stars form a class of

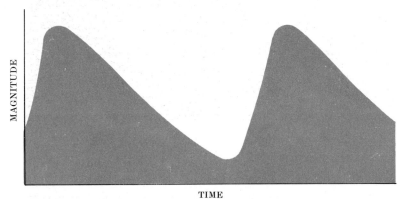

A typical light curve for a pulsating variable star

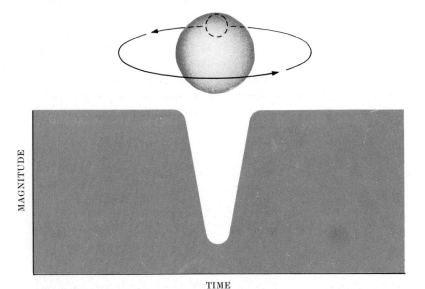

The light curve of an eclipsing binary system

semi-regular variables, the members of which are somewhat unpredictable. Finally, there are the irregular variables that have to be monitored regularly, as their behaviour cannot be predicted.

Novae and supernovae

Totally unexpected 'new' stars are called novae or supernovae, depending on type. A nova is an *old* star that suddenly increases in brightness by 10 to 15 magnitudes and then dies back again after a few weeks. This unpredictable burst of energy may bring a very faint star up to naked eye visibility so that a 'new star' appears temporarily in the heavens. Naked-eye novae occur every few years on average.

An even larger change is found in supernovae, which shine for a few days with the light of a thousand million suns. The supernova that gave rise to the Crab Nebula (M1) remained visible in daylight for several weeks. No supernova has been sighted visually in the Milky Way since the sixteenth century, but they are occasionally seen in other galaxies.

Double stars

A great many stars, perhaps one-quarter in all, are members of binary star systems, in which two stars orbit each other. When the orbit of a binary star is precisely edge-on as seen from the Earth, the members of the pair will regularly eclipse each other. This phenomenon results in the eclipsing variable stars, such as Algol (β Persei) which is in eclipse every 2 days 13 hours. The eclipse lasts for about seven hours, during which the brightness declines from 2.2 to 3.5 then climbs back to 2.2. magnitude.

Some apparently double stars are actually pairs of stars that are unrelated physically, but happen to be very close together as viewed from Earth. This chance lining-up of stars at different distances may easily be seen for Mizar, the second star of the bear's tail (or plough's handle) in Ursa Major. Mizar (ζ Ursae Majoris) has a faint companion star, Alcor (80 Ursae Majoris), which is 12 minutes of arc (one-third of the Moon's apparent diameter) away. Mizar and Alcor are not physically related, but each of them is, in turn, a multiple star – Mizar has three component stars and Alcor two, although these cannot be distinguished in a small telescope.

The study of stars that vary is a fascinating undertaking for amateur astronomers, and one in which skilled observers can assist professional researchers. In many countries the national societies for amateur astronomers coordinate the work of variable star observers.

In order to keep this atlas relatively simple, the variable stars are present on the maps but are not marked in any distinctive manner. Some of the brightest ones are noted in the table and each may easily be found on the appropriate map. When a variable star is being observed, its magnitude should be estimated by comparing its apparent brightness to that of constant stars in the same field of view. It should be possible to make an estimate correct to 0.5 magnitude; a trained observer can achieve estimates correct to 0.1 magnitude without difficulty.

Use of instruments

Field glasses

If you are a beginner, field glasses will enable you to see the rich star fields of the Milky Way, star clusters, and bright nebulae, for little cost. But it will not be possible to distinguish planetary detail or structure in the fainter nebulae. Before purchasing a telescope capable of doing this, seek advice from a local or national society of amateur astronomers.

The optical characteristics to consider when choosing field glasses are linear magnification (generally 7 to 10 times) and lens aperture (generally 30–60mm). Any magnification higher than about 10 shows hand-tremble so much that a rigid mount (e.g. a camera tripod) is essential. A larger aperture is better for astronomical viewing because it captures more of the precious faint light from celestial objects. The ideal combination for elementary astronomy is probably 10×50 (magnification $\times$ aperture), but this size may feel a little heavy to some observers. For specialist work, such as comet seeking, a specification of 20×80 might be preferred. Whatever type of field glasses is acquired, it is essential that the lens be fully-coated and that the images do not suffer from false colours. It is also important that there be some means of adjusting the internal prisms, should these become misaligned.

Refracting telescopes

The least expensive refracting telescopes have objective lenses about 60mm in diameter. Such a telescope must have a good tripod, a range of eyepieces, controls for guiding the telescope by hand in order to counteract the rotation of the celestial sphere, and a small finder telescope, which is used for aiming the main telescope at the chosen fields of view. The eyepieces have their focal length marked on them; the linear magnification is found by dividing the telescope focal length by the eyepiece focal length. Thus a 25mm eyepiece will give a magnification $\times 40$ with a 1-metre (i.e. 1000mm) telescope. The highest usable magnification is usually around $\times 120$ with inexpensive instruments. Even with an instrument of professional specification it is unusual for magnifications exceeding $\times 300$ to produce any worthwhile images. This is because turbulence in the Earth's atmosphere causes the image to dance about, and no amount of magnification can restore clarity to an image that has been blurred by the atmosphere. The most crucial parts of the small telescope are the lenses: these must be colour-corrected.

Reflecting telescopes

Small reflecting telescopes have mirror diameters of 15 to 25cm. Because they use an aluminized, curved mirror to form the primary image, the problems of false colour that beset field glasses and refracting telescopes do not arise. It is possible to purchase basic reflecting telescopes that can be up-graded subsequently by obtaining accessories, such as electric drive and graduated circles that give a read-out of right ascension and declination.

Using the atlas

Crab nebula in Taurus, M1

When you want to see the stars and constellations it is important for your eyes to be protected from direct light. In the dark the pupil of the eye becomes fully open in about twenty minutes and the eye is then at its most sensitive. Any source of bright light is not only painful, but causes the pupil to close, destroying the dark adaptation. However, some light is necessary if the atlas is referred to while viewing. The best way of doing this is to fit a dim bulb to a flashlight or torch, and then to cover the flashlight with translucent red paper. This produces a soft, red glow – the eye is least affected by red light – that should adequately illuminate the atlas.

Maps 1 and 2 show the circumpolar stars visible on any night from high northern and high southern latitudes. To use Maps 3 to 8 select the map for which the tables give a local time and date closest to the time of observation. This map shows the stars on the observer's meridian.

The maps immediately adjacent show the stars more than two hours of right ascension (30 degrees) to the east and west of the meridian. You need to know your latitude to deduce the extreme declination visible (stars only on the horizon when on the meridian). In the northern hemisphere the extreme declination visible is $-(90° - \text{latitude N})$. In the southern hemisphere it is $+(90° - \text{latitude S})$. A good, rough basis for the declination limit for the novice observer is $-30°$ for much of Europe and North America, and $+60°$ for Australasia and South Africa.

Map 1
North Celestial Pole
+60° to +90° declination

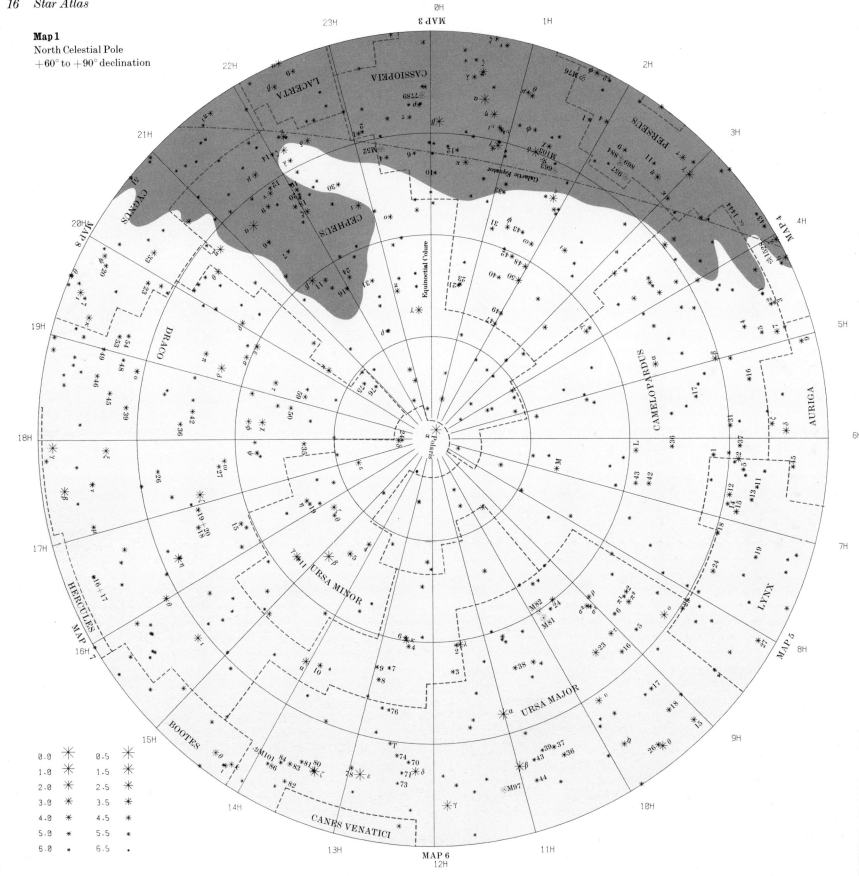

MAP 6
12H

Map 2
South Celestial Pole
−60° to −90° declination

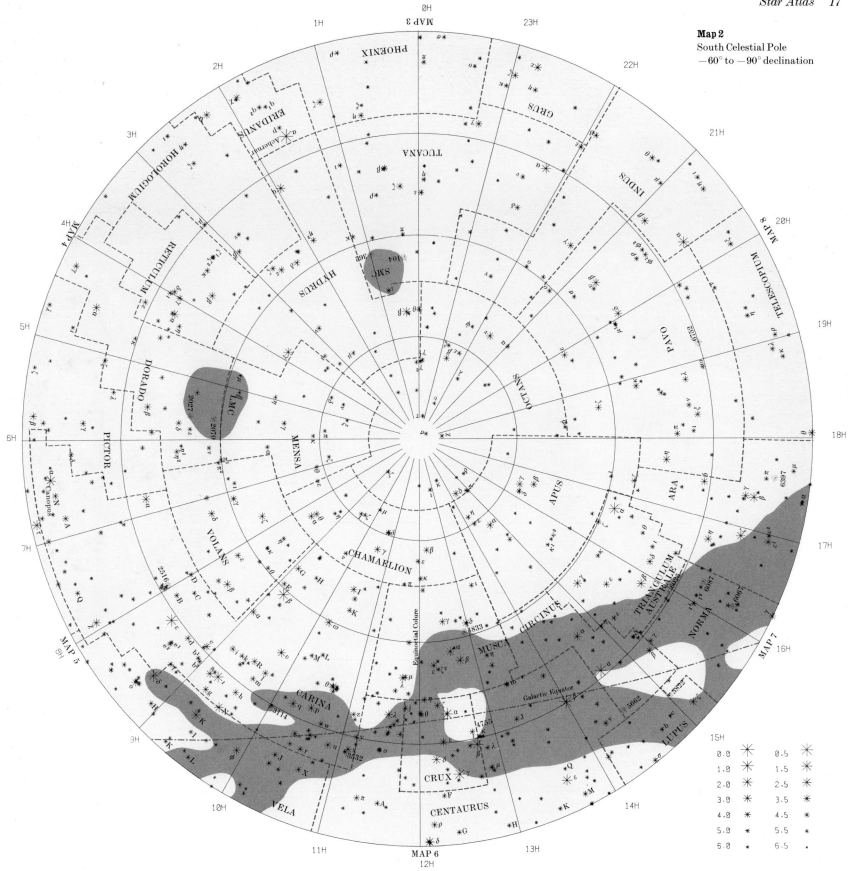

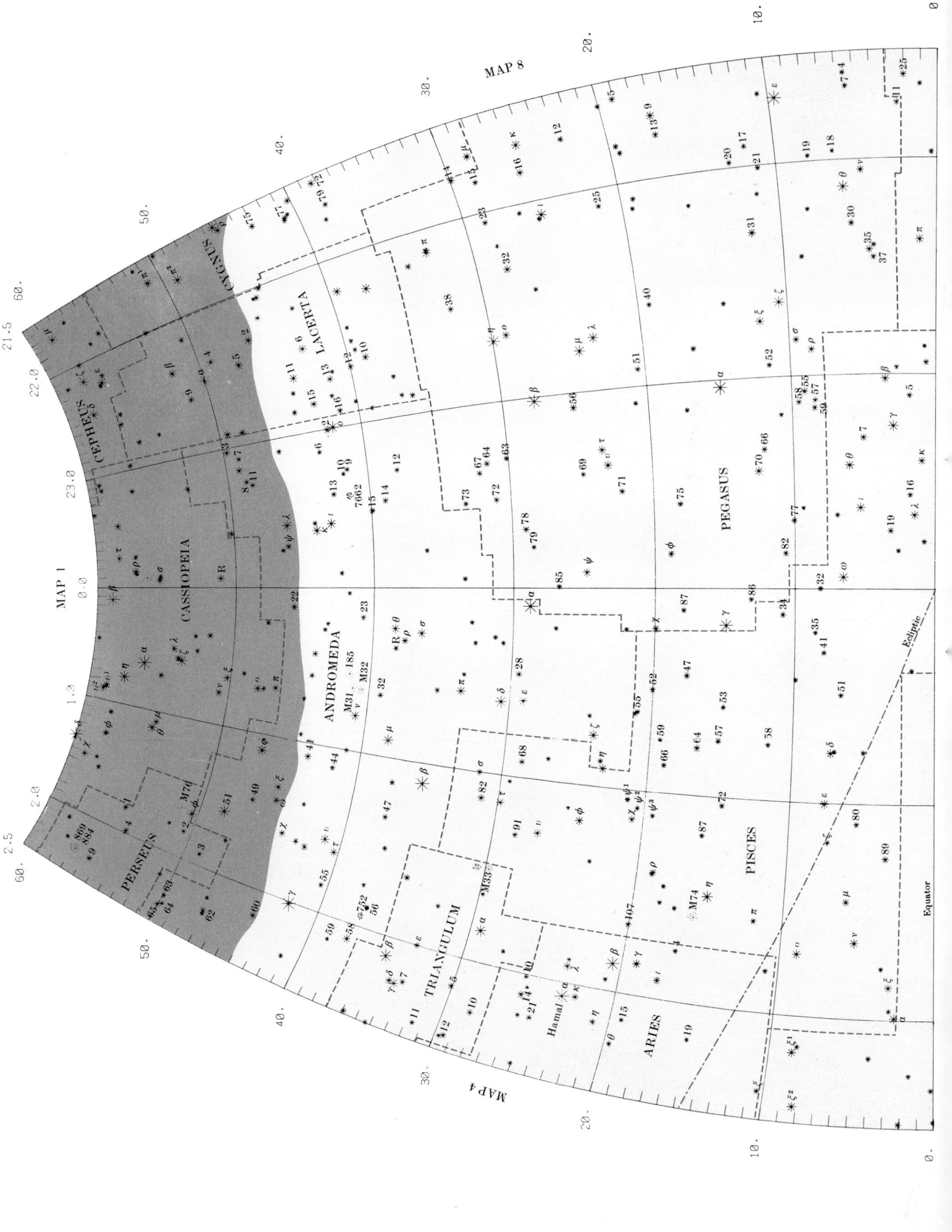

Map 3
0 hours Right Ascension lies on the meridian on

7 July	05.00
6 August	03.00
6 September	01.00
6 October	23.00
6 November	21.00
6 December	19.00
5 January	17.00

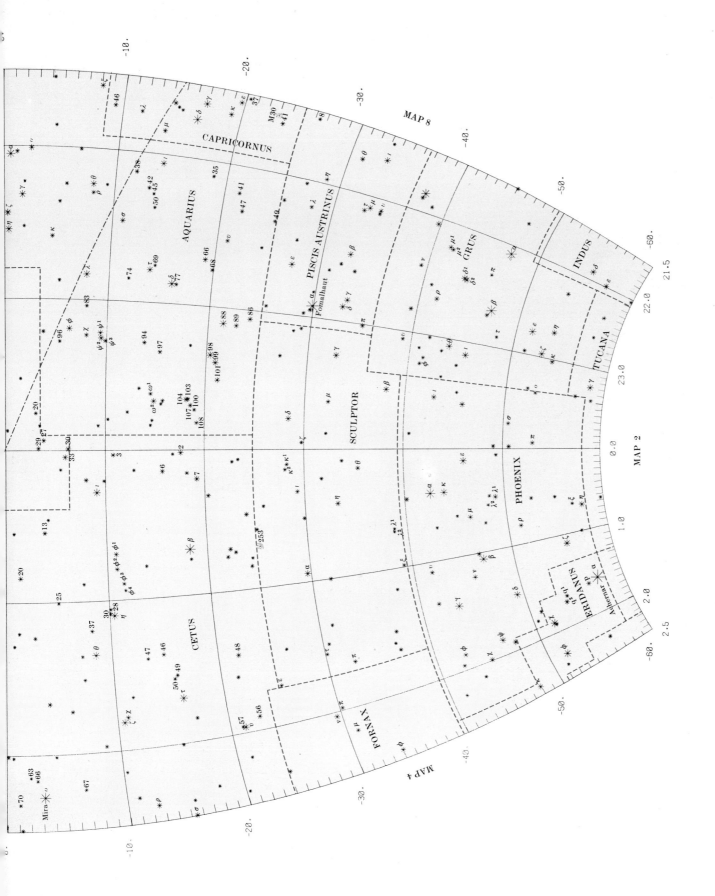

Map 4

4 hours Right Ascension lies on the meridian on

6 September	05.00
6 October	03.00
6 November	01.00
6 December	23.00
5 January	21.00
5 February	19.00
7 March	17.00

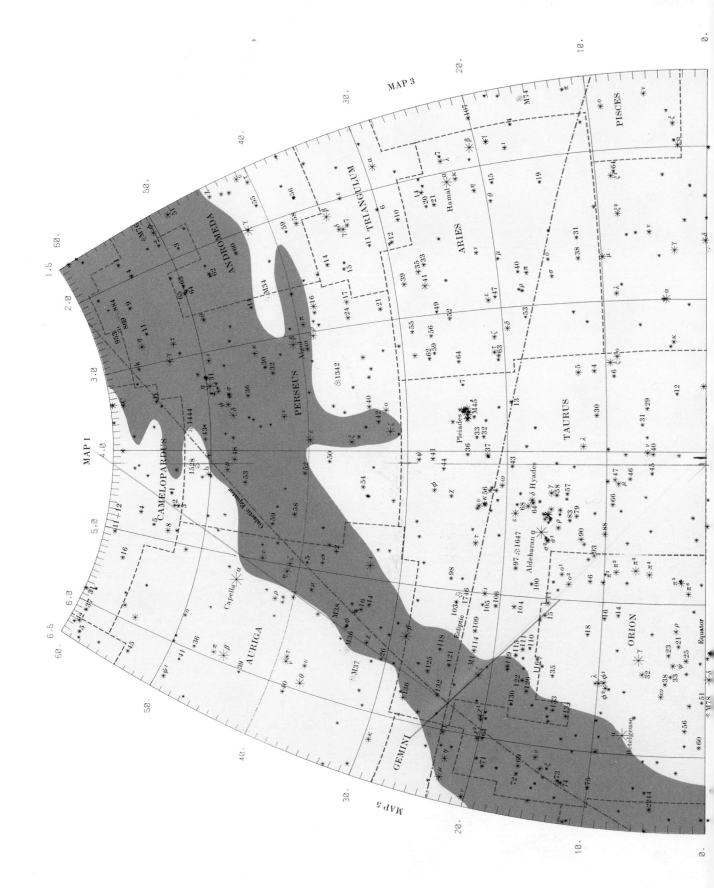

MAP 3

MAP 1

MAP 5

PISCES

ARIES

TRIANGULUM

ANDROMEDA

PERSEUS

CAMELOPARDUS

AURIGA

TAURUS

ORION

GEMINI

Hamal

Pleiades

Hyades

Aldebaran

Capella

Betelgeuse

Algol

Ecliptic

Equator

(Galactic Equator)

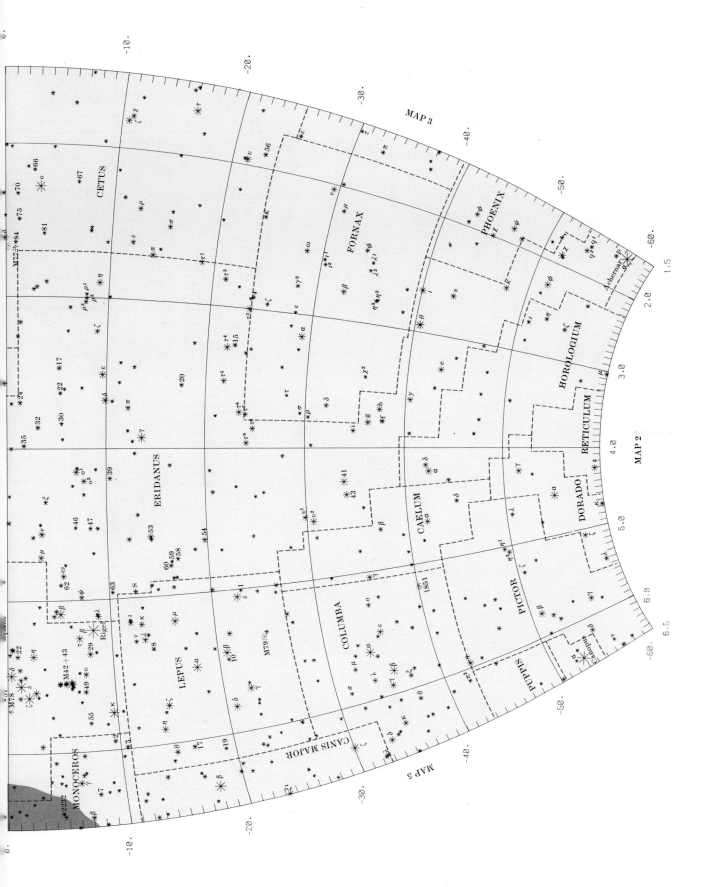

Map 5
8 hours Right Ascension lies on the meridian on

6 November	05.00
6 December	03.00
5 January	01.00
5 February	23.00
7 March	21.00
6 April	19.00
7 May	17.00

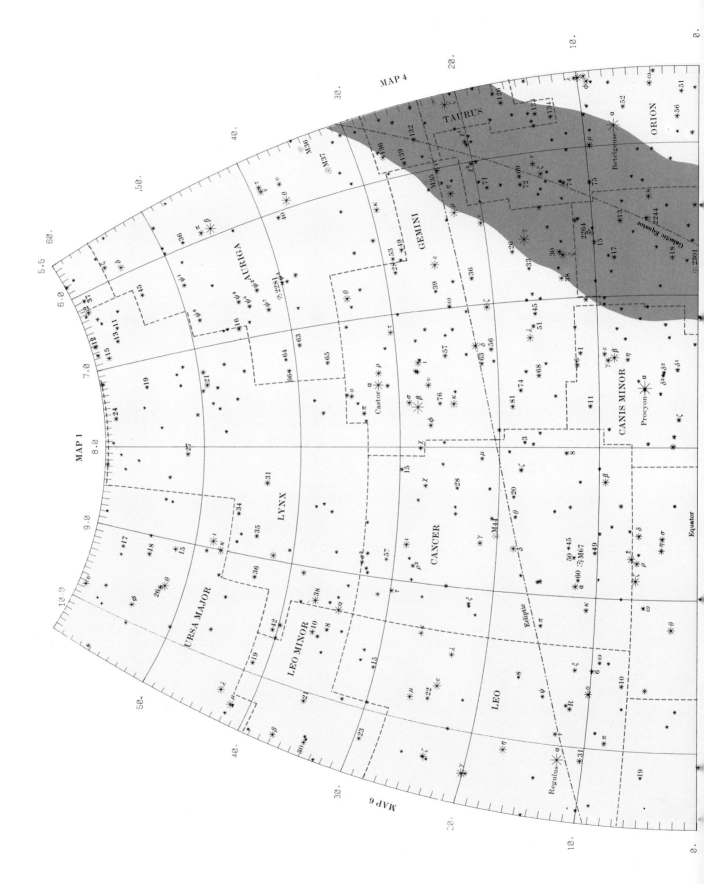

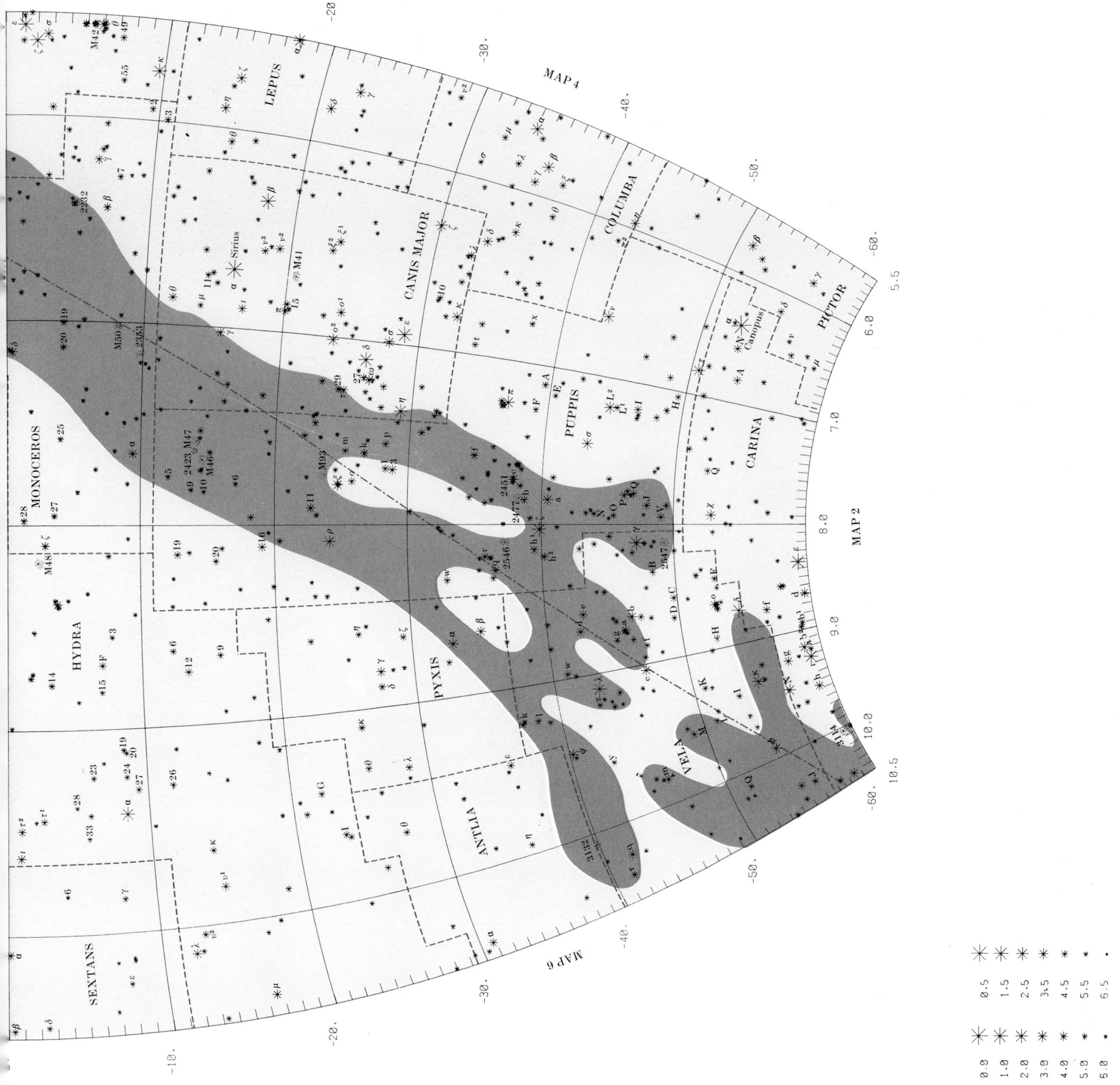

MAP 4

MAP 2

MAP 6

LEPUS

CANIS MAJOR

COLUMBA

PICTOR

CARINA

PUPPIS

MONOCEROS

HYDRA

PYXIS

VELA

ANTLIA

SEXTANS

Sirius

Canopus

-10.

-20.

-30.

-40.

-50.

-60.

5.5

6.0

7.0

8.0

9.0

10.0

10.5

0.5	1.5	2.5	3.5	4.5	5.5	6.5

0.0	1.0	2.0	3.0	4.0	5.0	6.0

Map 6
12 hours Right Ascension lies on the meridian on

5 January	05.00
5 February	03.00
6 March	01.00
6 April	23.00
7 May	21.00
6 June	19.00
7 July	17.00

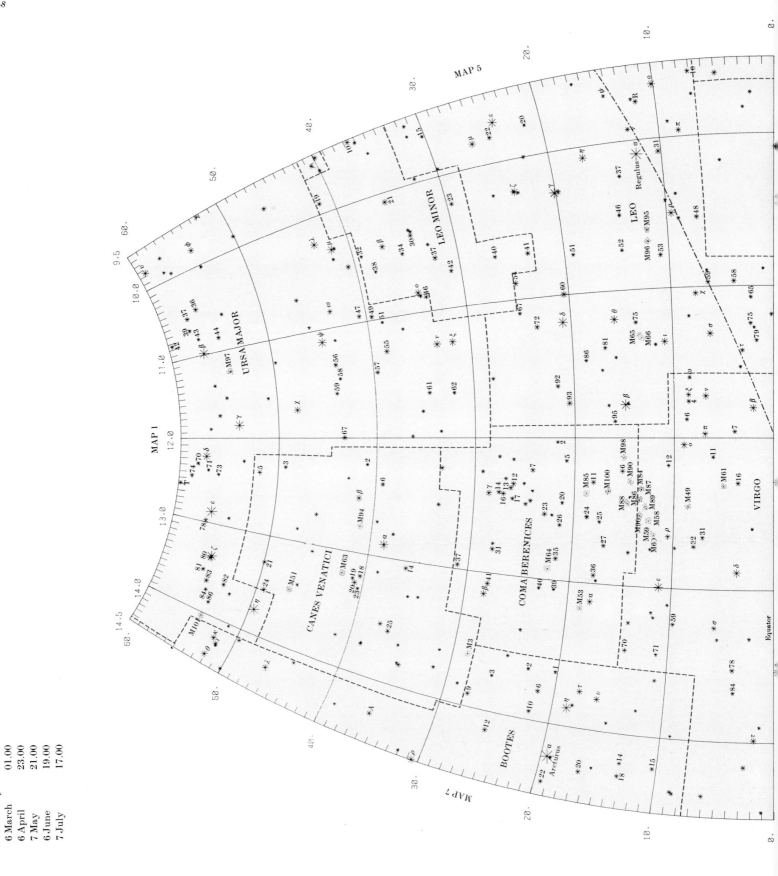

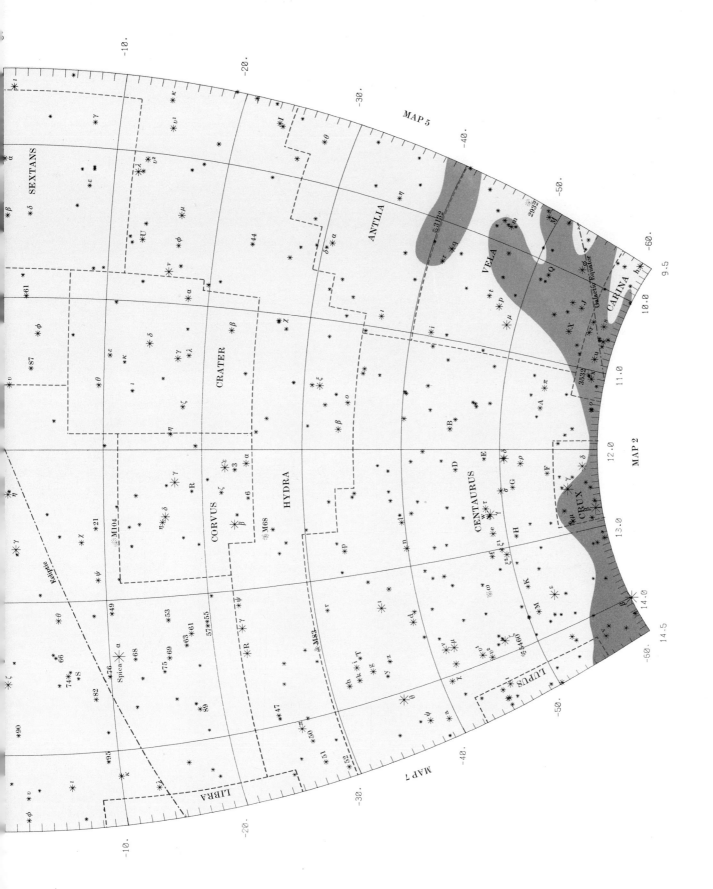

MAP 5

SEXTANS

ANTLIA

VELA

CARINA

CRATER

HYDRA

CORVUS

CENTAURUS

CRUX

LUPUS

LIBRA

MAP 7

MAP 2

Spica

M104

M68

Ecliptic

Galactic Equator

0.0	0.5
1.0	1.5
2.0	2.5
3.0	3.5
4.0	4.5
5.0	5.5
6.0	6.5

Map 7
16 hours Right Ascension lies on the meridian on

7 March	05.00
6 April	03.00
7 May	01.00
6 June	23.00
7 July	21.00
6 August	19.00
6 September	17.00

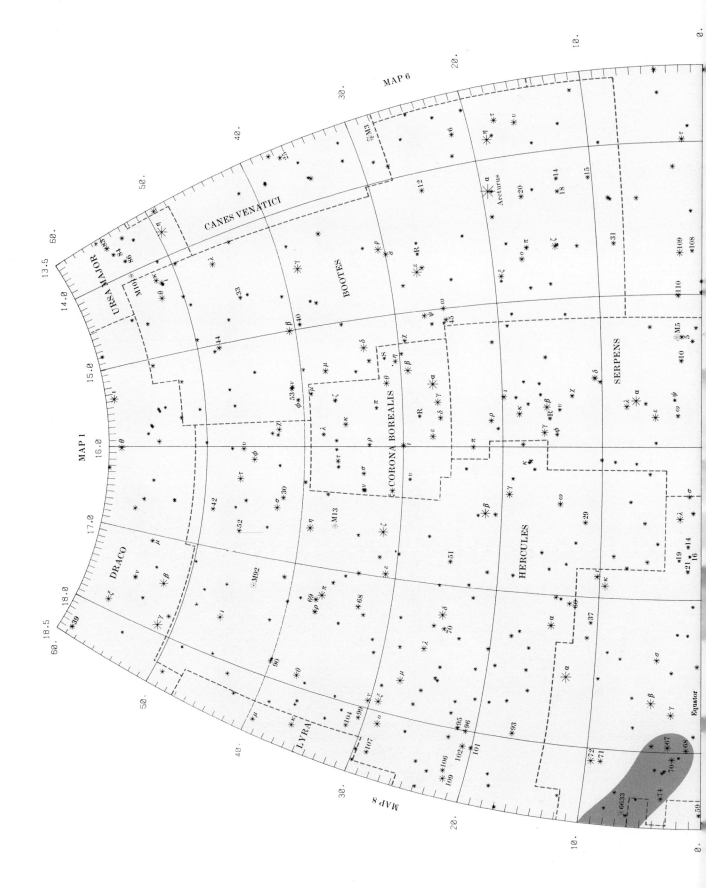

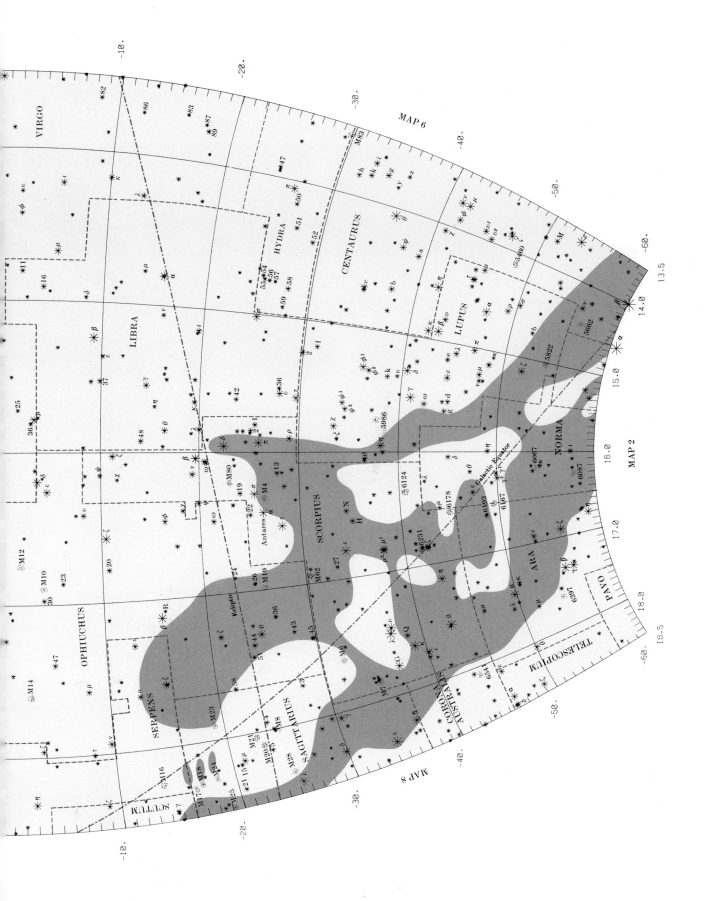

MAP 6

MAP 2

MAP 8

VIRGO

LIBRA

HYDRA

CENTAURUS

LUPUS

NORMA

OPHIUCHUS

SERPENS

SCORPIUS

ARA

PAVO

TELESCOPIUM

CORONA AUSTRALIS

SAGITTARIUS

SCUTUM

Antares

Galactic Equator

Ecliptic

✳	✳	✳
0.5	1.5	2.5

✳ 3.5
✳ 4.5
* 5.5
. 6.5

✳ 0.0
✳ 1.0
✳ 2.0
✳ 3.0
* 4.0
* 5.0
. 6.0

Map 8
20 hours Right Ascension lies on the meridian on

7 May	05.00
6 June	03.00
7 July	01.00
6 August	23.00
6 September	21.00
6 October	19.00
6 November	17.00

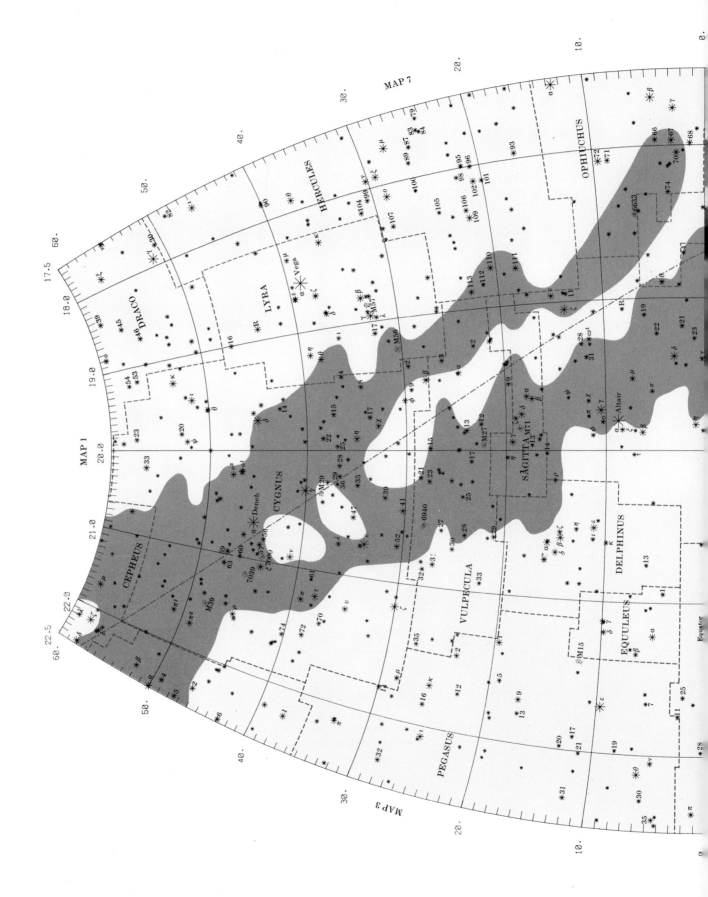

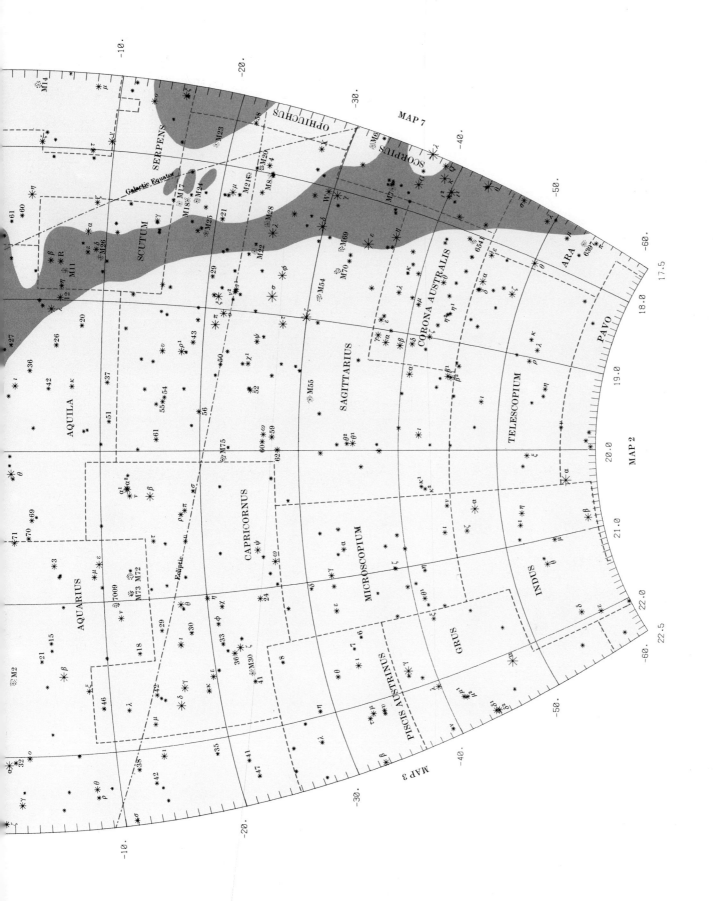

List of variable stars

Some variable stars that can be seen with the naked eye at least some of the time

Name	Right Ascension h m	Declination ° ′	Type	Magnitude range	Period days	Map
γ Cas	0 05	+60 00	irregular	1.6– 3.2	—	1
ζ Phe	1 06	−55 31	eclipsing	3.6– 4.1	1.67	3
o Cet	2 17	−03 12	long period	1.7–10.1	331	4
ρ Per	3 02	+38 39	semi-regular	3.3– 4.2	33–55	4
β Per	3 05	+40 46	eclipsing	2.2– 3.5	2.87	4
λ Tau	3 58	+12 21	eclipsing	3.3– 4.2	3.95	4
α Ori	5 53	+07 24	semi-regular	0.2– 1.0	2070 ?	4
ζ Gem	7 01	+20 39	Cepheid	3.7– 4.3	10.2	5
L² Pup	7 12	−44 33	semi-regular	2.6– 6.0	140.8	5
R Car	9 31	−62 34	long period	3.9–10.0	380.6	2
U Hya	10 35	−13 07	irregular	4.8– 5.8	450 ?	6
R Hya	13 27	−23 01	long period	4.0–10.0	386	6
δ Lib	14 58	−08 19	eclipsing	4.8– 6.1	2.3	7
R CrB	15 47	+28 19	irregular	5.8–15	—	7
α Her	17 12	+14 27	irregular	3.0– 4.0	—	7
β Lyr	18 48	+33 18	eclipsing	3.4– 4.1	12.9	8
κ Pav	18 52	−67 18	Cepheid	4.0– 5.5	9.1	2
χ Cyg	19 49	+32 47	long period	3.3–14.2	406.9	8
η Aqu	19 50	+00 53	Cepheid	3.7– 4.7	7.2	8
μ Cep	21 42	+58 33	irregular	3.6– 5.1	—	8
δ Cep	22 27	+58 10	Cepheid	3.6– 4.3	5.37	3
β Peg	23 01	+27 49	semi-regular	2.1– 3.0	2.40	3
ρ Cas	23 52	+57 13	irregular	4.1– 6.2	—	3

List of constellations

Name	Genitive	Abbreviation	English Equivalent	Map
Andromeda	Andromedae	And	Daughter of Cepheus	3, 4
Antila	Antliae	Ant	Air Pump	5, 6
Apus	Apodis	Aps	Bird of Paradise	2
Aquarius	Aquarii	Aqr	Water Bearer	3, 8
Aquila	Aquilae	Aql	Eagle	8
Ara	Arae	Ara	Altar	2, 7
Aries	Arietis	Ari	Ram	3, 4
Auriga	Aurigae	Aur	Charioteer	4, 5
Bootes	Bootis	Boo	Bear Driver	6, 7
Caelum	Caeli	Cae	Sculptor's Chisel	4
Camelopardus	Camelopardalis	Cam	Giraffe	1
Cancer	Cancri	Cnc	Crab	5
Canes Venatici	Canum Venaticorum	CVn	Hunting Dogs	6
Canis Major	Canis Majoris	CMa	Greater Dog	5
Canis Minor	Canis Minoris	CMi	Lesser Dog	5
Capricornus	Capricorni	Cap	Goat	8
Carina	Carinae	Car	Keel	2
Cassiopeia	Cassiopeiae	Cas	Cassiopeia	1, 3
Centaurus	Centauri	Cen	Centaur	2, 6, 7
Cepheus	Cephei	Cep	Cepheus	1
Cetus	Ceti	Cet	Sea Monster	3, 4
Chamaeleon	Chamaeleontis	Cha	Chameleon	2
Circinus	Circini	Cir	Compasses	2
Columba	Columbae	Col	Dove	4, 5
Coma Berenices	Comae Berenices	Com	Berenice's Hair	6
Corona Australis	Coronae Australi	CrA	Southern Crown	8
Corona Borealis	Coronae Borealis	CrB	Northern Crown	7
Corvus	Corvi	Crv	Crow or Raven	6
Crater	Crateris	Crt	Cup	6
Crux	Crucis	Cru	Southern Cross	2

List of constellations

Name	Genitive	Abbreviation	English Equivalent	Map
Cygnus	Cygni	Cyg	Swan	8
Delphinus	Delphini	Del	Dolphin	8
Dorado	Doradus	Dor	Swordfish	2
Draco	Draconis	Dra	Dragon	1, 8
Equuleus	Equulei	Equ	Foal	8
Eridanus	Eridani	Eri	River	3, 4
Fornax	Fornaci	For	Furnace	3, 4
Gemini	Geminorum	Gem	Twins	5
Grus	Gruis	Gru	Crane	3, 8
Hercules	Herculis	Her	Hercules	7, 8
Horologium	Horologii	Hor	Clock	2, 4
Hydra	Hydrae	Hya	Water Serpent	5, 6, 7
Hydrus	Hydri	Hyi	Water Snake	2
Indus	Indi	Ind	American Indian	2, 8
Lacerta	Lacertae	Lac	Lizard	3
Leo	Leonis	Leo	Lion	5, 6
Leo Minor	Leonis Minoris	LMi	Lion Cub	5, 6
Lepus	Leporis	Lep	Hare	4, 5
Libra	Librae	Lib	Scales or Balance	7
Lupus	Lupi	Lup	Wolf	7
Lynx	Lyncis	Lyn	Lynx	1, 5
Lyra	Lyrae	Lyr	Lyre	8
Mensa	Mensae	Men	Table Mountain	2
Microscopium	Microscopii	Mic	Microscope	8
Monoceros	Monocerotis	Mon	Unicorn	5
Musca	Muscae	Mus	Fly	2
Norma	Normae	Nor	Carpenter's Square	7
Octans	Octantis	Oct	Octant	2
Ophiuchus	Ophiuchi	Oph	Serpent Holder	7, 8
Orion	Orionis	Ori	Great Hunter	4, 5
Pavo	Pavonis	Pav	Peacock	2
Pegasus	Pegasi	Peg	Winged Horse	3, 8
Perseus	Persei	Per	Perseus	3, 4
Phoenix	Phoenicis	Phe	Phoenix	3, 4
Pictor	Pictoris	Pic	Painter's Easel	2, 4
Pisces	Piscium	Psc	Fishes	3
Piscis Austrinus	Piscis Austrini	PsA	Southern Fish	3, 8
Puppis	Puppis	Pup	Stern	5
Pyxis	Pyxidis	Pyx	Compass Box	5
Reticulum	Reticuli	Ret	Net	2
Sagitta	Sagittae	Sge	Arrow	8
Sagittarius	Sagittarii	Sgr	Archer	7, 8
Scorpius	Scorpii	Sco	Scorpion	7
Sculptor	Sculptoris	Scl	Sculptor's Workshop	3
Scutum	Scuti	Sct	Shield	8
Serpens	Serpentis	Ser	Serpent	7, 8
Sextans	Sextantis	Sex	Sextant	5, 6
Taurus	Tauri	Tau	Bull	4
Telescopium	Telescopii	Tel	Telescope	8
Triangulum	Trianguli	Tri	Triangle	3, 4
Triangulum Australe	Trianguli Australe	TrA	Southern Triangle	2
Tucana	Tucanae	Tuc	Toucan	2
Ursa Major	Ursae Majoris	UMa	Greater Bear	1, 5, 6
Ursa Minor	Ursae Minoris	UMi	Lesser Bear	1
Vela	Velae	Vel	Sail	5, 6
Virgo	Virginis	Vir	Virgin	6, 7
Volans	Volantis	Vol	Flying Fish	2
Vulpecula	Vulpeculae	Vul	Fox	8

List of nebulae

NGC	Messier, etc.		Right Ascension h m	Declination ° ′	Magni- tude	Type of nebula	Map
104	47 Tuc	Tuc	0 23	−72 13	4.5	globular cluster	2
224	M31	And	0 41	+41 08	4.5	spiral galazy	3
253		Scl	0 46	−25 26	7.5	spiral galaxy	3
581	M103	Cas	1 31	+60 35	7.0	open cluster	1
598	M33	Tri	1 33	+30 32	7.0	spiral galaxy	3
752		And	1 56	+37 33	6.5	open cluster	3
869	χ Per	Per	2 17	+57 02	4.5	open clusters, together	4
884	h Per	Per	2 20	+57 00	4.5	forming famous double cluster, h and χ Persei	4
1034	M34	Per	2 40	+42 40	6.0	open cluster	4
1342		Per	3 30	+37 15	7.0	open cluster	4
	M45	Tau	3 44	+24	1.5	'Pleiades', open cluster	4
1444		Per	3 48	+52 35	6.5	open cluster	4
1528		Per	4 13	+51 11	6.5	open cluster	4
	Hyades	Tau	4 20	+17		open cluster	4
1647		Tau	4 45	+19 02	6.0	open cluster	4
1746		Tau	5 02	+23 47	6.0	open cluster	4
1904	M79	Lep	5 23	−23 33	8.5	globular cluster	4
1912	M38	Aur	5 27	+35 49	7.0	open cluster	4
1976	M42	Ori	5 34	− 5 24		'Great Nebula' – diffuse emission nebula	4
1960	M36	Aur	5 35	+34 07	6.5	open cluster	4
2099	M37	Aur	5 51	+32 32	6.0	open cluster	4
2168	M35	Gem	6 07	+24 21	5.5	open cluster	5
2232		Mon	6 25	− 4 43	4.0	open cluster	5
2244		Mon	6 30	+ 4 54	5.0	open cluster	5
2264		Mon	6 40	+ 9 55	4.0	open cluster	5
2287	M41	CMa	6 46	−20 44	5.0	open cluster	5
2281		Aur	6 47	+41 05	7.0	open cluster	5
2301		Mon	6 50	+ 0 30	6.5	open cluster	5
2323	M50	Mon	7 00	− 8 16	6.5	open cluster	5
2422	M47	Pup	7 35	−14 25	4.5	open cluster	5
2423		CMa	7 36	−13 48	7.0	open cluster	5
2437	M46	Pup	7 41	−14 46	6.5	open cluster	5
2447	M93	Pup	7 44	−23 49	6.5	open cluster	5
2451		Pup	7 45	−37 55	3.5	open cluster	5
2477		Pup	7 51	−38 29	5.5	open cluster	5
2516		Car	7 58	−60 48	3.5	open cluster	2
2547		Vel	8 10	−49 11	5.0	open cluster	5
2548	M48	Hya	8 11	− 5 38	5.5	open cluster	5
2632	M44	Cnc	8 39	+20 05	4.0	'Praesepe', open cluster	5
2682	M67	Cnc	8 50	+11 54	7.5	open cluster	5
3034	M82	UMa	9 53	+69 48	9.5	irregular galaxy	1
3031	M81	UMa	9 54	+69 11	8.5	spiral galaxy	1
3132		Ant	10 06	−40 18	8.0	planetary nebula	6
4755	κ Cru	Cru	12 52	−60 13	5.0	'Jewel Box', open cluster	2
5024	M53	Com	13 11	+18 18	8.5	globular cluster	6
5139	ω Cen	Cen	13 25	−47 11	4.0	globular cluster	6
5194	M51	CVn	13 29	+47 19	10.0	'Whirlpool', spiral galaxy	6
5272	M3	CVn	13 41	+28 30	7.0	globular cluster	6
5457	M101	UMa	14 02	+54 29	8.5	spiral galaxy	7
5460		Cen	14 06	−48 12	6.0	open cluster	7
5822		Lup	15 03	−54 15	6.5	open cluster	7
5904	M5	Ser	15 17	+ 2 11	7.0	globular cluster	7
6025		TrA	16 02	−60 26	6.0	open cluster	2
6067		Nor	16 09	−54 05	6.5	open cluster	7
6087		Nor	16 16	−57 51	6.0	open cluster	7
6093	M80	Sco	16 16	−22 56	8.5	globular cluster	7

List of nebulae

List of nebulae

NGC	Messier, etc.		Right Ascension h m	Declination ° ′	Magni- tude	Type of nebula	Map
6121	M4	Sco	16 22	−26 27	7.5	globular cluster	7
6193		Ara	16 40	−48 43	5.5	open cluster	7
6205	M13	Her	16 40	+36 33	6.0	globular cluster	7
6218	M12	Oph	16 46	− 1 55	8.0	globular cluster	7
6254	M10	Oph	16 56	− 4 04	7.5	globular cluster	7
6266	M62	Sco	17 00	−30 05	8.0	globular cluster	7
6273	M19	Oph	17 01	−26 13	8.5	globular cluster	7
6341	M92	Her	17 16	+43 12	6.0	globular cluster	7
6402	M14	Oph	17 36	− 3 16	9.5	globular cluster	7
6405	M6	Sco	17 38	−32 12	4.5	open cluster	7
6475	M7	Sco	17 52	−34 48	3.5	open cluster	7
6514	M20	Sgr	18 01	−23 02	5.0	'Trifid', emission nebula	8
6523	M8	Sgr	18 02	−24 23	5.0	'Lagoon', emission nebula	8
6531	M21	Sgr	18 03	−22 30	7.0	open cluster	8
6611	M16	Ser	18 15	+36 04	6.5	open cluster	8
6603	M24	Sgr	18 17	−18 25	6.0	open cluster	8
6613	M18	Sgr	18 19	−17 08	8.0	open cluster	8
6618	M17	Sgr	18 19	−16 11	7.0	'Omega', emission nebula	8
6633		Oph	18 26	+ 6 33	5.5	open cluster	8
	M25	Sgr	18 30	−19 16	6.0	open cluster	8
6637	M69	Sgr	18 30	−32 22	9.0	globular cluster	8
6656	M22	Sgr	18 35	−23 57	6.5	globular cluster	8
6681	M70	Sgr	18 42	−32 20	9.0	globular cluster	8
6705	M11	Scu	18 50	− 6 18	7.0	'Wild Duck', open cluster	8
6720	M57	Lyr	18 52	+33 00	9.5	'Ring', planetary nebula	8
6715	M54	Sgr	18 54	−30 30	8.5	globular cluster	8
6752		Pav	19 09	−60 02	7.0	globular cluster	8
6809	M55	Sgr	19 39	−31 00	7.0	globular cluster	8
6838	M71	Sge	19 53	+18 43	8.5	globular cluster	8
6853	M27	Vul	19 59	+22 39	7.5	'Dumbbell', planetary nebula	8
6864	M75	Sgr	20 05	−22 00	9.5	globular cluster	8
6940		Vul	20 34	+28 13	6.5	open cluster	8
7009		Aqr	21 03	−11 28	8.5	'Saturn', planetary nebula	8
7078	M15	Peg	21 29	+12 04	7.5	globular cluster	8
7092	M39	Cyg	21 31	+48 20	5.5	open cluster	8
7089	M2	Aqr	21 32	− 0 56	7.5	globular cluster	8
7099	M30	Cap	21 39	−23 18	8.5	globular cluster	8
7654	M52	Cas	23 23	+61 27	8.0	open cluster	1
7662		And	23 25	+42 24	9.0	planetary nebula	3

Overleaf: The 'Lagoon' Nebula, M8

Acknowledgements

The Publishers gratefully acknowledge permission to reproduce
the following illustrations:
Hale Observatories 6, 7, 10 top right, 13 left, top right, centre
right; *Lick Observatory* 8, 9 left, 11, 12 bottom; *Royal
Astronomical Society* 4, 9 right, 10 left, centre right, 12 top, centre,
13 bottom right, 15; *Royal Observatory Edinburgh* 10 bottom
right

The Star Atlas was computed from copyrighted data kindly
supplied by the Smithsonian Astrophysical Observatory,
Cambridge, Massachusetts, USA on the IBM 370/165 computer
at the University of Cambridge, UK

Computer programmer: Patricia Stewart

The diagrams are by Michael Robinson